FLEXIBLE PAVEMENT MAINTENANCE OPERATIONS

Dr. Faiq M. S. Al-Zwainy
Assistant Professor
Al-Nahrain University, Iraq

Dr. Ibraheem A. Aidan
Assistant Professor
Al Maarif University College, Iraq

lulu

Lulu Press, Inc.

Faiq M. S. Al-Zwainy

Ibraheem A. Aidan

Publisher's note
Every possible effort has been made to ensure that the information contained in this book is accurate at the time of going to press, and the publishers and authors cannot accept responsibility for any errors or omissions, however caused. No responsibility for loss or damage occasioned to any person acting, or refraining from action, as a result of the material in this publication can be accepted by the editor, the publisher or any of the authors.

First published in the United States in 2018 by Lulu Press.

Library of Congress Cataloging-in-Publication Data
Application Submitted
British Library Cataloguing-in-Publication Data
A catalogue record for this book is available from the British Library.

ISBN 978-0-359-02797-2

Printed and bound in United States by Lulu Press

Faiq M. S. Al-Zwainy

Ibraheem A. Aidan

DEDICATION

To Iraq

CONTENTS

ACKNOWLEDGMENTS

Thanks to Allah for providing with power and patience to achieve this work.

We would like to thank the Ministry of Higher Education and Scientific Research, in Republic of Iraq, for giving our chance for experiences this book. Moreover, it is our great pleasure to thank the College of Engineering in Al-Nahrain University that embraced ours during years of write this book.

We would like to express our gratitude to the many people who provided support, talked things over, read, wrote, offered comments, allowed our to quote their remarks and assisted in the editing, proofreading and design.

Grateful thanks and appreciation to our families for their continuous support to perform this book. Last and not least: I beg forgiveness of all those who have been with our over the course of the years and whose names I have failed to mention.

Authors

1 HIGHWAY PROJECT

1.1 Introduction

For as long as the human race has existed, transportation has played a significant role by facilitating commerce, trade, conquest, and social interaction, while consuming a considerable portion of resources and time. The primary need for transportation has been economic, involving personal travel in search of food or work, travel for the exchange of goods and commodities, exploration, personal fulfillment, and the improvement of a society or a nation. The traveling of people and goods, which is the basis of transportation, always has been considered to accomplish those objectives or tasks that require transfer from one location to another. The history of transportation illustrates that the way people move is affected by technology, cost, and demand. Transportation is a significant element in our daily life, also it represent about 18 percent of household expenditure and forming about 10 percent of the workforce[1].

1.2 Highways Importance

Highways have a vitally important to the development of country's economic. The construction of road network with a high quality directly increases a nation's economic output by reducing journey times and costs, and making a region more attractive economically. The actual construction process will have the added effect of stimulating the construction market

1.3 Highway Location Selection

Selecting the location of a proposed highway is a necessary initial step in its design. The decision to select a particular location is usually depend on topography of the region, environmental factors such as noise, soil characteristics, air pollution, and economic factors. The data required for the decision process are usually obtained from different types of surveys, depending on many factors being considered. Most engineering consultants and state agencies presently involved in highway locations use computerized techniques to process the vast amounts of data that are generally handled in the decision process. These techniques include remote sensing, which uses aerial photographs for the preparation of maps, and computer graphics, which is a combination of the analysis of computer-generated data with a display on a computer monitor. The highway location should provide the easy flow of traffic at the design capacity, while meeting design criteria and safety standards. The highway should also cause a minimal disruption to historic and archeological sites and to other land-use activities. Environmental impact studies are therefore required in most cases before a highway location is finally agreed upon.

The highway location process involves four phases:

1) Office study of existing information
2) Reconnaissance survey
3) Preliminary location survey
4) Final location survey

1.3.1. Office Study of Existing Information

The first phase in any highway location study is the examination of all available data of the area in which the road is to be constructed. This phase is usually carried out in the office prior to any field or photogrammetric investigation. All of the available data are collected and examined. These data can be obtained from existing engineering reports, maps, aerial photographs, and charts, which are usually available at one or more of the state's departments of transportation, agriculture, geology, hydrology, and mining. The type and amount of data collected and examined depend on the type of highway being considered, but in general, data should be obtained on the following characteristics of the area:

1) Engineering, including topography, geology, climate, and traffic volumes
2) Social and demographic, including land use and zoning patterns
3) Environmental, including types of wildlife; location of recreational, historic, and archeological sites; and the possible effects of air, noise, and water pollution
4) Economic, including unit costs for construction and the trend of agricultural, commercial, and industrial activities

1.3.2 Reconnaissance Survey

The object of this phase of the study is to identify several feasible routes, each within a band of a limited width of a few hundred feet. When rural roads are needed to build, there is often little information available on maps or photographs, and therefore aerial photography is widely used to obtain the required information. Feasible routes are identified by a stereoscopic examination of the aerial photographs, taking into consideration factors such as:

1) Terrain and soil conditions
2) Serviceability of route to industrial and population areas
3) Crossing of other transportation facilities, such as rivers, railroads, and highways
4) Directness of route

1.3.3 Preliminary Location Survey

In this phase preliminary alignments are used to evaluate the economic and environmental feasibility of the alternative routes. Where the purpose of economic evaluation is to evaluate each alternative route is carried out to determine the future effect of investing the resources necessary to construct the highway. And the Construction of a highway at any location will have a significant impact on its surroundings so the environmental Evaluation is significant. A highway is therefore an integral part of the local environment and must be considered as such. This environment includes plant, animal, and human communities and encompasses social, physical, natural, and man-made variables. These variables are interrelated in a manner that maintains equilibrium and sustains the lifestyle of the different communities.

1.3.4 Final Location Survey

The final location survey is a detailed layout of the selected route. The horizontal and vertical alignments are determined, the positions of structures, and the drainage channels are located.

1.4 Factors Influencing Highway Design

Highway design is based on specified design standards and controls which depend on the following roadway system factors:

1) Functional classification

2) Design hourly traffic volume and vehicle mix

3) Design speed

4) Design vehicle

5) Cross section of the highway, such as lanes, shoulders, and medians

6) Presence of heavy vehicles on steep grades

7) Topography of the area that the highway traverses

8) Level of service

9) Available funds

10) Safety

11) Social and environmental factors

These factors are often interrelated. For example, design speed depends on functional classification which is usually related to expected traffic volume. The design speed may also depend on the topography, particularly in cases where limited funds are available. In most instances, the principal factors used to determine the standards to which a particular highway will be designed are the level of service to be provided, expected traffic volume, design speed, and the design vehicle. These factors, coupled with the basic characteristics of the driver, vehicle, and road, are used to determine standards for the geometric characteristics of the highway, such as cross sections and horizontal and vertical alignments.

1.5 Highway Strata

A highway pavement is composed of a system of overlaid strata of chosen processed materials that is positioned on the in-situ soil, termed the sub-grade. Its basic requirement is the provision of a uniform skid-resistant running surface with adequate life and requiring minimum maintenance. The chief structural purpose of the pavement is the support of vehicle wheel loads applied to the carriageway and the distribution of them to the sub-grade immediately underneath. The pavement designer must develop the most economical combination of layers that will guarantee adequate dispersion of the incident wheel stresses so that each layer in the pavement does not become overstressed during the design life of the highway·

The major variables in the design of a highway layers are:

1) The thickness of each layer in the pavement.
2) The material contained within each layer of the pavement.
3) The type of vehicles in the traffic stream.
4) The volume of traffic predicted to use the highway over its design life.
5) The strength of the underlying sub-grade soil.

There are three basic layers of the highway pavement which they are listed below:

a) Foundation: The foundation consists of the native sub-grade soil and the layer of graded stone (sub-base and possibly capping) immediately overlaying it. The function of the sub-base and capping is to provide a platform on which to place the road-base material as well as to insulate the sub-grade below it against the effects of inclement weather. These layers may form the temporary road surface used during the construction phase of the highway.

b) Road-base: The road-base is the main structural layer whose main function is to withstand the applied wheel stresses and strains incident on it and distribute them in such a manner that the materials beneath it do not become overloaded.

c) Surfacing: The surfacing combines good riding quality with adequate skidding resistance, while also minimizing the probability of water infiltrating the pavement with consequent surface cracks. Texture and durability are vital requirements of a good pavement surface as are surface regularity and flexibility. For flexible pavements, the surfacing is normally applied in two layers – base-course and wearing course – with the base-course an extension of the road-base layer but providing a regulating course on which the final layer is applied. In the case of rigid pavements, the structural function of both the road-base and surfacing layers are integrated within the concrete slab.

1.6 Highway Drainage

Provision of sufficient drainage is an important factor in the location and geometric design of highways. Drainage facilities on any highway or street should adequately provide for the flow of water away from the surface of the pavement to properly designed channels. Inadequate drainage will eventually result in serious damage to the highway structure. In addition, traffic may be slowed by accumulated water on the pavement, and accidents may occur as a result of hydroplaning and loss of visibility from splash and spray. The importance of enough drainage is recognized in the amount of highway construction dollars allocated to drainage facilities.

About 25 percent of highway construction dollars are spent for erosion control and drainage structures, such as culverts, bridges, channels, and ditches. The highway engineer is concerned primarily with two sources of water. The first, surface water, is that which occurs as rain or snow. Some of this is absorbed into the soil, and the remainder remains on the surface of the ground and should be removed from the highway pavement. Drainage for this source of water is referred to as surface drainage. The second source, ground water, is that which flows in underground streams. This may become important in highway cuts or at locations where a high water table exists near the pavement structure. Drainage for this source is referred to as subsurface drainage·

1.6.1 Highway Drainage Structures

Drainage structures are constructed to carry traffic over natural waterways that flow below the right of way of the highway. These structures also provide for the flow of water below the highway, along the natural channel, without significant alteration or disturbance to its normal course. One of the main concerns of the highway engineer is to provide an adequate size structure, such that the waterway opening is sufficiently large to discharge the expected flow of water. Inadequately sized structures can result in water impounding, which may lead to failure of the adjacent sections of the highway due to embankments being submerged in water for long periods. The two general categories of drainage structures are major and minor. Major structures are those with clear spans greater than 20 ft, whereas minor structures are those with clear spans of 20 ft or less. Major structures are usually large bridges, although multiple-span culverts also may be included in this class. Minor structures include small bridges and culverts .

1.6.1.1 Major Structures

Emphasis is placed on selecting the span and vertical clearance requirements for such structures. The bridge deck should be located above the high water mark. The clearance above the high water mark depends on whether the waterway is navigable. If the waterway is navigable, the clearance above the high water mark should allow the largest ship using the channel to pass underneath the bridge without colliding with the bridge deck. The clearance height, type, and spacing of piers also depend on the probability of ice jams and the extent to which floating logs and debris appear on the waterway during high water.

1.6.1.2 Minor Structure

Minor structures, consisting of short-span bridges and culverts, are the predominant type of drainage structures on highways. Although openings for these structures are not designed to be adequate for the worst flood conditions, they should be large enough to accommodate the flow conditions that might occur during the normal life expectancy of the structure. Provision also should be made for preventing clogging of the structure due to floating debris and large boulders rolling from the banks of steep channels. Culverts are made of different materials and in different shapes. Materials used to construct culverts include concrete (reinforced and unreinforced), corrugated steel, and corrugated aluminum.

1.7 Designations Associated with Road Projects

While the researcher was reading the references related to the highway engineering, many names related to this type of projects were noticed. Examples of these names are highway, freeway, street, road, and others, so the difference between one name and other should be investigated from point of view of the researcher. These differences are described below:

1) A beltway: (also called a loop) circles around a town or a city.

2) A causeway: is a roadway built on top of raised land made of compacted soil usually to cross a body of water.

3) A driveway: privately maintained access road that leads to a single house or a group of homes.

4) A freeway: there are no traffic lights or intersections to slow traffic down. Traffic is usually divided using a median which are concrete barriers or sometimes grass.

5) A highway: refers to any public road that is well-maintained and can handle a heavy load of traffic.

6) The Interstate: The Interstate Highway System is a complex network of freeways that connect states together.

7) A parkway: is a major throughway originally designed for scenic and recreational driving.

8) A roundabout: is a circular intersection where traffic is designed to flow without stopping, in one direction.

9) A street: connects buildings together, while providing vehicles and pedestrians access. The distinction between *road* and *street* is that a road serves primarily vehicles and sometimes lacks a sidewalk, whereas streets are designed for pedestrian travel *and* vehicles.

1.8 Highway Project Development Process

This process can follow a six-step that which include the following phases:

1) **Select Project:** Highway projects are selected based on a variety of criteria. These include: public concerns, traffic crash data,

pavement and bridge condition, traffic volume and trends, and forecasts of future growth.

2) **Investigate Alternates:** After a project is selected for inclusion, the alternatives are identified and each alternative is analyzed and assessed based on cost and its impact on people, businesses, farmlands, wetlands, endangered species, historic structures, artifacts, and landfills.

3) **Obtain Final Approvals:** Information about acquisitions from farm operations is furnished to the Department of Agriculture. In other words, the projects that are state-funded should be approved by the Department of Transportation.

4) **Develop Project Design:** The specific project route and related details are finalized. Affected property owners are contacted to discuss land purchases and relocation plans. Every effort is made to ensure that offering prices reflect "just compensation" for the property. Local businesses affected by the project receive assistance during the construction phase intended to mitigate the economic effect of the project. Among the services that may be stipulated during construction are: (1) maintaining access for customers, employees, and service vehicles, (2) posting highway signs that inform motorists of business district locations, (3) permitting the temporary posting of signs in the highway right-of-way to reassure customers that businesses are accessible, and (4) assisting businesses to identify and inform motorists of alternate routes during construction. Access management is analyzed to ensure that the existing highway system continues to perform with acceptable efficiency and safety. The following engineering documents are prepared: (1) A plan which depicts the physical layout of the

project, (2) specifications which define how each item in the plan is to be built, and (3) engineering estimates that list the expected cost of each item of work.

5) **Prepare for Construction:** the owner firm of Highway Construction reviews each plan, specification, and estimate package and prepares a document suitable to be used by a contractor in preparing a bid. All land required for the project must have been purchased so that the project site can be prepared for construction. The construction schedule is coordinated with utility companies in the event that replacement of sewer, gas, power, or phone lines are required. Citizens receive prior notification should there be a need to disrupt utility service. Request for project bids are advertised and those received within a specified period are checked for completeness and accuracy. All bids awarded are forwarded to the Governor for signature. With contractor contracts approved and signed.

6) **Construct the Project**: A pre-construction meeting is held with the contractor, local utilities, Department of Natural Resources, and local government officials. The DOT coordinates with property owners and local businesses to ensure that all prior commitments to landowners, such as access to homes and businesses, are fulfilled. They also meet with the contractors throughout the construction period. Citizens are kept informed of construction progress through meetings and news releases sent to local media. When the project has completed the new roadway is then operated to ensure safety and traffic flow and will be inspected every two years to monitor its pavement performance.

1.9 Types of Pavements

There are two main types of pavements which they rigid and flexible pavement also they will be next totally described.

1.9.1 Rigid Pavements

A rigid pavement consists of a sub-grade/sub-base foundation covered by a slab constructed of pavement quality concrete. The concrete slab must be of sufficient depth so as to prevent the traffic load causing premature failure. Appropriate measures should also be taken to prevent damage due to other causes. The proportions within the concrete mix will determine both its strength and its resistance to climate changes and general wear. The required slab dimensions are of great importance this can be founded by the design procedure. Joints in the concrete may be formed in order to aid the resistance to tensile and compressive forces set up in the slab due to shrinkage effects. The concrete may be reinforced with steel [20]. The following diagram could help in clarifying the layers of this pavement type which addressed :

1.9.1.1 Types of Rigid Highway Pavements

Rigid highway pavements can be divided into three general types: plain concrete pavements, simply reinforced concrete pavements, and continuously reinforced concrete pavements. The definition of each pavement type is related to the amount of reinforcement used. The three types are listed below and totally described:

1.9.1.1.1 Jointed Plain Concrete Pavement

Plain concrete pavement has no temperature steel or dowels for load transfer. However, steel tie bars often are used to provide a hinge effect at longitudinal joints and to prevent the opening of these joints. Plain concrete pavements are used mainly on low-volume highways or when cement-stabilized soils are used as sub-base. Joints are placed at relatively shorter distances (10 to 20 ft) than with other types of concrete pavements to reduce the amount of cracking. So it is necessary, to know the joints types and the description of each type, this will be displayed in the next section. The researcher thinks that it is necessary to give a description to the joints and their types so they next will be listed.

1.9.1.1.1.1 Joints Types in Concrete Pavements

Joints are provided in a pavement slab in order to allow for movement caused by changes in moisture content and slab temperature. Transverse joints across the pavement at right angles to its centre line permit the release of shrinkage and temperature stresses. The greatest effect of these stresses is in the longitudinal direction. Longitudinal joints, on the other hand, deal with induced stresses most evident across the width of the pavement. There are four main types of transverse joints:

1) Contraction joints: allow induced stresses to be released by permitting the adjacent slab to contract, thereby causing a reduction in tensile stresses within the slab. The joint, therefore, must open in order to permit this movement while at the same time prohibiting vertical movement between adjacent concrete slabs.

2) Expansion joints: the full discontinuity exists between the two sides, with a compressible filler material included to permit the

adjacent concrete to expand. These can also function as contraction or warping joints.

3) Warping joints: are required in plain unreinforced concrete slabs only. They permit small angular movements to occur between adjacent concrete slabs. Warping stresses are very likely to occur in long narrow slabs. They are required in unreinforced slabs only.

4) Construction joint: Construction is normally organized so that work on any given day ends at the location of an intended contraction or expansion joint. Where this proves not to be possible, a construction joint can be used. No relative movement is permitted across the joint

1.9.1.1.2 Simply Reinforced Concrete Pavement

Simply reinforced concrete pavements have dowels for the transferring of traffic loads across joints, with these joints spaced at larger distances, ranging from 30 to 100 ft. Temperature steel is used throughout the slab, with the amount dependent on the length of the slab. Tie bars also are used commonly at longitudinal joints.

1.9.1.1.3 Continuously Reinforced Concrete Pavement

Continuously reinforced concrete pavements have no transverse joints, except construction joints or expansion joints when they are necessary at specific positions, such as at bridges. These pavements have a relatively high percentage of steel, with the minimum usually at 0.6 percent of the cross section of the slab. They also contain tie bars across the longitudinal joints

1.9.2 Flexible Pavements

The surfacing and road-base materials, bound with bitumen binder, overlay granular unbound or cement-bound material. As it is known, the flexible pavement consisting of many layers which they are:

1) The sub-grade: is usually the natural material located along the horizontal alignment of the pavement and serves as the foundation of the pavement structure. It also may consist of a layer of selected borrow materials, well compacted to get the needed specifications. It may be necessary to treat the sub-grade material to achieve certain strength properties required for the type of pavement being constructed.

2) The sub-base: Located immediately above the sub-grade, the sub-base component consists of material of a superior quality to that which is generally used for sub-grade construction. The requirements for sub-base materials usually are given in terms of the gradation, plastic characteristics, and strength. In cases where suitable sub-base material is not readily available, the available material can be treated with other materials to achieve the necessary properties. This process of treating soils to improve their engineering properties is known as stabilization.

3) The base: The base course lies immediately above the sub-base. It is placed immediately above the sub-grade if a sub-base course is not used. This course usually consists of granular materials such as crushed stone, crushed or uncrushed slag, crushed or uncrushed gravel, and sand. The specifications for base course materials usually include more strict requirements than those for sub-base materials, particularly with respect to their plasticity, gradation, and strength.

4) The wearing surface: it is the upper course of the road pavement and constructed immediately above the base course. The surface course in flexible pavements usually consists of a mixture of mineral aggregates and asphalt. It should be capable of withstanding high tire pressures, resisting abrasive forces due to traffic, providing a skid-resistant driving surface, and preventing the penetration of surface water into the underlying layers. The thickness of the wearing surface can vary from 3 in. to more than 6 in., depending on the expected traffic on the pavement. The following diagram could help in clarifying the layers of this pavement type which addressed :

1.10 Pavement Management

Pavement management is a systematic process for maintaining, upgrading, and operating physical pavement assets in a cost-effective manner. The process combines applications of established engineering principles with sound business practices and economic theory, thus assuring an organized and scientific approach to decision making. The process involves the following steps for a given pavement section: (1) assess present pavement condition, (2) predict future condition, (3) conduct an alternatives analysis, and (4) select an appropriate rehabilitation strategy [19]. The pavement management process is strategic and seeks answers to questions such as:

1) What is the current condition of the pavement network?
2) What resources in time, material and personnel will be required to maintain the network at a specified performance standard?
3) What should be the annual work program to address the most critical needs that reflect available resources?

1.11 Pavement Deterioration

As previously mentioned there are two types of pavements so that the deterioration in one type will be different from another. So, each type deterioration will be separately described in the next sections.

1.11.1 Deterioration in Flexible Pavements

For heavily trafficked roads, experience has indicated that, deterioration in the form of cracking/deformation is most likely to be occurred in the surface of the pavement rather than deeper down within its structure. A well-constructed pavement will have an extended life span on condition that distress, seen in the form of surface cracks and ruts, is taken care of before it starts to affect the structural integrity of the highway [19]. There are four basic phases of structural deterioration for a flexible pavement which they listed below:

1) Phase 1: When a new/strengthened pavement is reaching stability, at which point its load spreading ability is still improving.

2) Phase 2: Load spreading ability is quite even and the rate of structural deterioration can be calculated with some confidence.

3) Phase 3: At this stage structural deterioration becomes less predictable and strength may decrease gradually or even rapidly. This is the 'investigatory' phase. A pavement entering this phase should be monitored in order to ascertain what if any remedial action is required to be carried out on it.

4) Phase 4: Here the pavement has deteriorated to failure. Strengthening can only be achieved by total reconstruction. This phase can last quite a number of years, with maintenance becoming necessary with increasing frequency until the point is reached where

the costs associated with this treatment make reconstruction the cheaper option.

It is truly to say that, the remedial work during the third 'investigatory' phase is more economic than total reconstruction at the end of its full design life.

1.11.2 Deterioration in Rigid Pavements

Cracking in rigid concrete slabs can be promoted by stresses generated at the edge/corner of slabs. These can vary from narrower hairline cracks which often appear while concrete is drying out, to 'wide' cracks (>0.15 cm) which result in the effective loss of aggregate interlock, allowing water to enter its structure and cause further deterioration. 'Medium' cracking greater than 0.5 mm will result in partial loss of aggregate interlock. Failure is defined as having occurred in an unreinforced concrete pavement if one of the following defects is present.

1) A medium or wide crack crossing the bay of the concrete slab longitudinally or transversely
2) A medium longitudinal and medium transverse crack intersecting, both exceeding 200 mm in length and starting from the edge of the pavement
3) A wide corner crack, more than 200 mm in radius, centered on the corner

1.12 Highway Pavement Material

It is known for all that there are many types of raw materials that used in highway projects, so each material will be briefly discussed in the following next sections.

1.12.1 Soils at Sub-Formation Level

Unless the subsoil is composed of rock, it is unlikely to be strong enough to carry even construction traffic. Therefore it is necessary to superimpose additional layers of material in order to reduce the stresses incident on it due to traffic loading. The in-situ soil would suffer permanent deformation if subjected to the high stresses arising from heavy vehicle traffic loading. The shear strength and stiffness modulus are accepted indicators of the susceptibility of the soil to permanent deformation. A soil with high values of both these characteristics will be less susceptible to permanent deformation. Both are usually reduced by increases in moisture content. Knowledge of them is essential within the pavement design process in order to determine the required thickness of the pavement layers. Since it is not always feasible to establish these two parameters for a soil, the California bearing ratio (CBR) test (will be discussed in the next section) is often used as an index test. While it is not a direct measure of either the stiffness modulus or the shear strength, it is a widely used indicator due to the level of knowledge and experience with it that has been developed by practitioners.

1.12.1.1 CBR Test

The CBR test acts as an attempt to quantify the behavioral characteristics of a soil trying to resist deformation when subject to a locally applied force such as a wheel load. Developed in California before World War II, to this day it forms the basis for the pre-eminent empirical pavement design methodology used that used everywhere. The test does not measure any fundamental strength characteristic of the soil. It involves a cylindrical plunger being driven into a soil at a standard rate of penetration, with the level of resistance of the soil to this penetrative effort being measured. The test can be done either on site or in the laboratory. If the test is done in the laboratory, it is important that the moisture content and dry density of the sample being tested should approximate as closely as possible those expected once the pavement is in place. All particles greater than 20 mm in diameter should first be removed. If done in situ, the test should be performed on a newly exposed soil surface at such a depth that seasonal variations in moisture content would not be expected.

1.12.2 Sub-Base

The sub-base and capping together act as a regulator of the surface of the sub-grade below and protect it against the effects of inclement weather. They, along with the sub-grade, provide a secure platform on which the upper layers of the highway pavement can be built. The determinant of the thickness of this section of the pavement is the strength of the underlying sub-grade. Its design is independent of the cumulative traffic incident on the upper layers of the pavement over its design life. For sub-grades in excess of 5% CBR, the required sub-base depth is no greater than 225 mm, down to a minimum of 150 mm at a sub-grade CBR of 15%. Granular and cement-based sub-bases are recommended for flexible pavements while only cemented sub-bases are recommended for rigid-type pavements.

The following five broad categories when sub-base layer is applied should be taken into account:

1) No sub-base layer is required if the sub-grade layer is consist of hard rock or of a granular material with a CBR value of at least 30%, also provided the water table is not at a high level.

2) In the case of sub-grades layer with a CBR greater than 15%, a sub-base thickness of 150 mm is necessary (in practical terms this constitutes the minimum sub-base layer thickness for ensuring proper spreading and compaction).

3) Where the CBR value of the sub-grade lies between 2.5% and 15%, two options are founded:

4) use 150 mm of sub-base layer over a layer of capping material, the thickness of which depends on the CBR of the sub-grade layer, or

5) When the layer of sub-base have CBR value greater than15% the layer thickness should be 150 mm at least and 350 mm (at 2.5% CBR) in thickness.

6) For all pavements where the sub-grade CBR is less than 2.5% and for rigid pavement construction on materials with CBR below 15%, 150 mm of sub-base must be used on top of capping. The thickness of the capping layer will reach 600 mm where the CBR of the sub-grade dips below 2%.

7) Where the sub-grade CBR value is substantially less than 2%, the material will often be removed and replaced with other more suitable material. The depth of this imported material would typically be between 500 mm and 1000 mm deep. Though this material may in reality be quite strong, it will be assumed to have a CBR of 2% and will thus require a 600 mm capping layer.

1.12.3 Materials within Flexible Pavements

The materials that used in one type of pavements are rather differs from another. So each type materials will be discussed separately.

1.12.3.1 Bitumen

Bitumen is produced artificially from crude oil within the petroleum refining process. It is a basic constituent of the upper layers in pavement construction. It can resist both deformation and changes in temperature. Its binding effect eliminates the loss of material from the surface of the pavement and prevents water penetrating the structure. Two basic types of bituminous binder exist.

1) Tar – obtained from the production of coal gas or the manufacture of coke

2) Bitumen – obtained from the oil refining process.

With the decreased availability of tar, bitumen is the most commonly used binding/water resisting material for highway pavements.

1.12.3.2 Asphalts

Two asphalts types will be discussed in this section: mastic asphalt and hot rolled asphalt:

1) Hot rolled asphalt: It is a dense material with low air voids content, consisting of a mixture of aggregate, fines, binder and a filler material, but in this case the grading is far less continuous (gap-graded) with a higher proportion of both fines and binder present in the mix. It is typically have from zero to 55% coarse aggregate

content, with base courses having either 50% or 60% and road-bases normally at 60%.

2) Mastic asphalt: is a very durable heavy-duty, weather-proof wearing course material. It consists of a mixture of asphaltic cement (low-penetration grade bitumen), fine aggregate and filler in proportions which result in a low-void impermeable mass. It contains a low percentage of coarse aggregate, all of which must pass the 14 mm sieve and be retained on the 10 mm sieve. The mix consists of a high percentage of fine aggregate, with no less than 45% and no more than 55% passing the 0.075 mm sieve and at least 97% passing the 2.36 mm sieve in addition to high proportions of both filler material and binder:

1.12.3.3 Aggregates

The maximum nominal aggregate size is determined from both the required thickness of the material when put in place and the surface texture called for. The size of aggregate must not be greater than the required layer thickness. The layer thickness must be approximately 21 /2 times the nominal maximum aggregate size, with a minimum thickness of 11 /2 times the nominal maximum aggregate size to minimize the likelihood of the larger stones being crushed during rolling.

1.12.4 Materials in Rigid Pavements

The main material that used in this type of pavements is the concrete instead of bitumen or asphalt that used in the flexible pavement type. Generally, some materials will be briefly discussed as next sections.

1.12.4.1 Slab Concrete

As the strength of concrete develops with time, its 28-day value is taken for specification purposes, though its strength at 7 days is often used as an initial guideline of the mix's ultimate strength. Pavement quality concrete generally has a 28-day characteristic strength of 40 N/mm², termed C40 concrete. Ordinary Portland cement (OPC) is commonly used. The cement content for C40 concrete should be a minimum of 320 kg/m³. Air content of up to 5% may be acceptable with a typical maximum water cement ratio of 0.5 for C40 concrete. The effects of temperature are such that a continuous concrete slab is likely to fail prematurely due to induced internal stresses rather than from excessive traffic loading. If the slab is reinforced, the effect of these induced stresses can be lessened by the addition of further reinforcement that increases the slab's ability to withstand them. This slab type is termed continuous reinforced concrete (CRC). Alternatively, dividing the pavement into a series of slabs and providing movement joints between these can permit the release and dissipation of induced stresses. This slab type is termed jointed reinforced concrete (JRC). If the slab is jointed and not reinforced, the slab type is termed unreinforced concrete (URC). It is necessary to know, if joints are employed, their type and location are important factors.

1.12.4.2 Reinforcement

Reinforcement can be in the form of a prefabricated mesh or a bar-mat. The function of the reinforcement is to limit the extent of surface cracking in order to maintain the particle interlock within the aggregate. In order to maximize its bond with the concrete within the slab, care must be taken to ensure that the steel is cleaned thoroughly before use. Because the purpose of the reinforcement is to minimize cracking, it should be placed near the upper surface of the pavement slab. A cover of approximately 60 mm is usually required, though this may be reduced slightly for thinner slabs. It is normally stopped approximately 125 mm from the edge of a slab, 100 mm from a longitudinal joint and 300 mm from any transverse joint [20].

2 HIGHWAY MAINTENANCE

`2.1 Introduction

The network of roads and streets that connects and serves cities, towns, and villages is one of the most widely used means of transportation. In Iraq, as in many other countries, a very high percent of all passenger travel is on roads and streets, and almost all of country's food and other goods are transported all or part of the way from the farm or factory to the store by way of roads and streets.

2.2 Roads types in Baghdad City:

Ministry of housing in Iraq classifies roads in Baghdad City according to the traffic volume, construction specifications and service purpose of the road, into four categories as follows:

2.2.1 Freeways and Express Ways:

They are main ways dedicated for vehicles with high speed design up to 120 km/hr characterized with high traffic volume up to 100000 vehicles per day on both sides. Moreover, they are long and without cross roads, such as (Mohammad Al-Qasim Road).

Whereas express ways, particularly the roads directly implemented, have many crossroads organized by traffic lights, such as (Salah Al-deen Al-Ayoubi road). These roads are characterized with the width of a single path, which is 3.75 m, and the width of the median, which is 3.0 to 8.0 m.

Moreover, they are without roadside and provided by safety fences and special path to stop vehicles in emergency cases.

2.2.2 Principal Distributors:

They are main roads that divide Baghdad City into many sectors. The essential function of these roads is to connect Baghdad sectors with each other, such as (Palestine street and Al-Rabeie street). These roads are characterized with high traffic volume up to 60000 vehicles daily on both sides. They permit the vehicles to move with speed designed up to 80 km/hr. Principal distributors are long roads in which the width of a single path is 3.5 m, the median is 3.0 to 5.5 m. An additional path to vehicles stopping may be added with a width of 2.5 m, as well as, a pedestrian pavement of 2.0 to 10.0 m width.

2.2.3 Secondary Distributors:

They are main roads dividing each sector of Baghdad City into main parts, such as (Al-Kifah and Abu-Nuas streets). Secondary distributors, from which local roads branch, are characterized with medium traffic density up to 40000 vehicles daily on both sides. Moreover, they permit the vehicles to move with a speed designed up to 60 km/hr. They differ from principal distributors in the sense they are shorter and similar in the width of the path which is 3.5 m and the median which is 2.5 m.

A path vehicles stopping vehicles may be added with a width of 2.5 m, as well as, a pedestrian pavement which is 5-10 m. Secondary distributors net length was 1060 km in (1984) in Baghdad City and it is supposed to be 1250 km at (2000) if the plane is full implemented.

2.2.4 Local Roads:

Local roads spread among buildings and houses, and they flow into secondary distributors. They are characterized with a limited traffic volume, which does not exceed 10000 vehicles daily on both sides. These roads allow the movements of vehicles with a speed designed up to 40 km/hr. Local roads are often short and the width of a single path is 3 m without a median. A path vehicles stopping may be added of 1.5 m width, in addition, to the roadside which width does not exceed 5 m.

2.3 Pavement Types in Baghdad City:

In Baghdad City, there are many kinds of roads, such as elevated roads (bridges) and underpasses (tunnels). Most of these roads are stretched on the earth surface they are of three kinds, which are flexible pavement, rigid pavement and composite pavement road.

2.3.1 Flexible Pavement:

This kind of pavement constitutes the majority of Baghdad roads. These roads are structurally composed of lower layer, which represents the natural ground or the soil filling after fixing and increasing its ability to bear loads. Then there is the subbase layer, which is composed of sand-gravel mixture material then, the asphalt base layer and after that the asphalt binder layer.

2.3.2 Rigid Pavement:

These roads represent a considerable part of main roads in Baghdad City, particularly the express ways. Such roads are structurally composed of reinforced cement concrete ground built on the natural fixed ground. Engineers use such kind of roads when the expected traffic volume is high and when the natural ground is not solid enough to bear loads when using the flexible paving or when the available aggregate is little. Such kinds of roads are not included in this study.

2.3.3 Composite Pavement:

They are rigid paving roads to which a layer or more of asphalt concrete is added.

2.3.4 Other Types:

A considerable number of elevated roads recently spread in Baghdad City, particularly through express way networks.

This kind of roads contains special structural element like beams, columns (poles), extending joints and supporting points, etc. In addition to bridges, in Baghdad City, there are several long and short tunnels. This kind of tunnels contains special structural element such as supporting walls and slabs in addition to special methods for disposing rainwater. Tunnel and bridges are not included in this study because the matter requires special care.

2.4 Common Damages of Baghdad City Roads:

Roads are designed according to different specifications owing to the differences in the environment, resources and the purpose. However, roads are more subject to damage due to change in friction and hitting powers between the vehicle's wheels and the surface of the road. Moreover, roads are exposed to nature factors, which play a damaging role, resulting from variation of temperature and humidity between day and night or from one season to another.

Damage in the roads comes as results of one or more of the following reasons:

1) Natural and weather factors.
2) Mistakes in the design.
3) Mistakes in the implementation methods.
4) Failure in the quality control of materials.
5) Periodic maintenance negligence.
6) Road misuse because of vehicles with axle loads exceeding the allowed limits.

It is worth mentioning that the statistics of bridges and roads general state, have proved that 74% of the damages in Iraqi roads, result from the excess of the axial loads, particularly the damage resulting from them, equals the axial loads above to power four.

The most important damages happen to asphalt surface, including the damages arising from distortion, disintegration, cracks of all types and the damages that lead to skid hazard as shown in Figure (2.1).

2.5 Road Maintenance Concept:

Maintenance can be, generally, defined as a mixture of any processes taken to keep certain material in service, or to get it back to its normal condition. Subsequently, maintenance keeps materials in a good condition and in service, so as to do its job as required.

Highway maintenance, is a program to preserve, and restore a system of roadways with its elements to its designed configuration. Road maintenance can be defined as a way for keeping the road always in a good condition, as when it was first built. Moreover, maintenance provides safety, and the appropriate and economical movement from one place to another [13].

Maintenance can also be defined as a way of getting the road back to the condition when it has been first built, or it can be defined as periodic activities working under the normal circumstances of environment and traffic, to keep the pavement in a condition almost similar to that when it was first constructed. Maintenance also aims of keeping the road in service for as longer period as possible, in addition to ensure the passing of vehicles safety in the road. Moreover, maintenance ensures a good investment of money spent in construction the road.

Subsequently, road maintenance must be focused in two essential sides as follows:

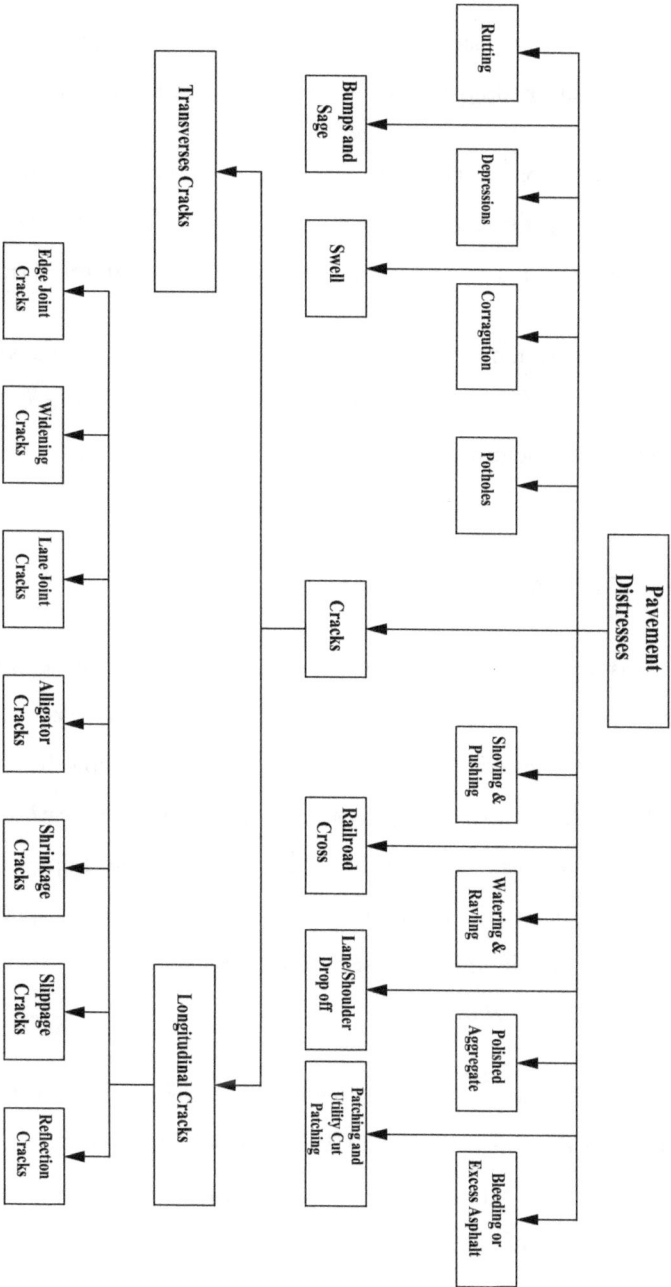

Figure 2.1   Types of asphalt pavement distresses.

First: - Road maintenance structure by keeping all the roads structural elements such as (asphalt layers, bas and sub bas layers, and shoulders) in a good condition almost similar to that when the roads were first constructed.

Second: - Road maintenance service, including maintenance concerning (traffic sign, protecting fences, wire fences, phones illumination poles, roadside, etc) in a way which ensures a safe and economical transport.

Since the maintenance activities are vary different from one state to another according to their circumstances and abilities, the following definitions are identified to indicate these activities as follows: -

(1) Preventive maintenance: -

It is a process of finding out the damages and repairing them when they first appear to prevent the damages before they become serious. The passing vehicles with loads, particularly those exceeding the designing loads, lead together with the bad weather conditions to increase and spread the damages within few days. The matter, which requires implementing detailed and continuous periodic examinations, is the walking of specialists on foot on paving or using highly observation devices.

(2) Rehabilitation:

They are activities through which the roads are restored to their original state after they disintegrate and damage where the periodic maintenance fails to repair them. Subsequently, these activities cause a substantive change in the paving constructive condition as adding asphalt layers and reconstruction.

Primary objectives of pavement rehabilitation, as indicated in the ASSHTO policy are to :

a)   Improving surface smoothness;

b)   Extending pavement life;

c)   Improving skid resistance;

d)   Reconstructing sections having poor foundations; and

e)   Improving drainage.

(3) Improvements:

These are activities of road parts reconstruction such as enlarging the horizontal curves or adding supported walls or paving shoulders in order to increase road safety, in addition to raising the service state. However, these activities do not increase the paving endurance force.

(4) Upgrading:

These are activities aim of increasing the paving endurance force, and may be accompanied by an increase in road's capacity (path number) in order to contain the new traffic volume and it's axial loads.

2.6 Maintenance Classification: -

Maintenance can be classified according to either the activities recurrent periods included there in or according to the activity type or their aim as shown in the following diagram:

```
                    Maintenance
                   Classification
```

Type of maintenance according to the activities type or their aim.	Type of maintenance according to the activities recurrent periods included there in

Structural maintenance	Routine maintenance
Safety maintenance	Recurrent maintenance
Amenity maintenance	Periodic maintenance
	Emergency maintenance

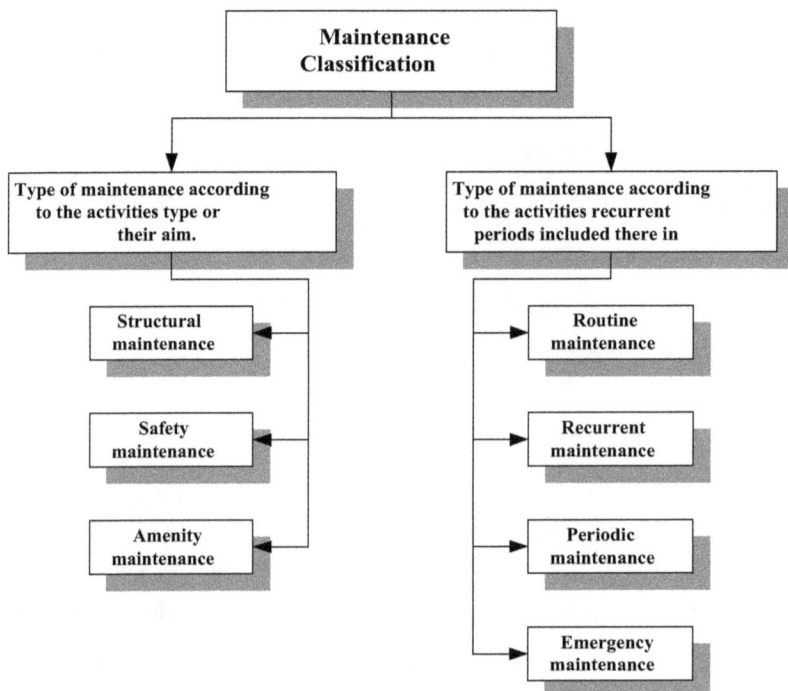

2.6.1 Routine Maintenance:

It is a group of activities that aim of keeping the road safety through daily examination by the road control officials on both sides, in order to ensure the road safety, in addition to take some immediate and protective measures, including the removal of all obstacles and the cleaning the road surface, the mediate, shoulders, and sides road.

Generally, the absence maintenance leads the road to more collapses, subsequently, the road loses a lot of its technical specifications and, as a result, it will be useless. In order to save a lot of amount of money, routine maintenance must be interested to prolong the rod's life and avoid the necessity for recurrent maintenance or reconstruction.

2.6.2 Recurrent Maintenance:

It is a group of activities that must be implemented on consecutive periods. Recurring this kind of maintenance depends on the traffic volume. An example of recurrent maintenance is repairing the crack .

2.6.3 Periodic Maintenance:

It is a number of activities that must be implemented in distant intervals to the extent of certain number of years, in order to repair the big damages that may occur in some parts of the road due to usage for a long period of time [17]. Implementing such maintenance requires necessary advanced preparations. An example of such maintenance is the addition of new asphalt layers or removing the damaged parts and reconstructing them after handling the reasons of damages or changing the pointed curves, or widening some roads.

Generally, the absence routine maintenance will lead to the implementation of periodic maintenance, which costs huge sums of money.

2.6.4 Urgent or Emergency Maintenance:

It is the kind of maintenance that must be implemented immediately and without any delay to repair any damage that may occur in the roads networks. Such as digging or the damages resulted from traffic accidents or the blocking of the road by sand hills or floods, etc.

Any delay in implementing such maintenance will cause loss of money, therefore, most of the countries pay great attention to provide well-trained and well-equipped urgent maintenance teams.

2.6.5 Structural Maintenance:

It is the group of activities that aims of maintaining the road structure including the paving surface, sewage and etc.

2.6.6 Safety Maintenance:

It is the group of activities that aims of maintaining parts of the road including illumination, safety fences, wire fences, traffic signs, etc. to ensure the safety of people using the road.

2.6.7 Amenity:

It is a group of activities that provide comfortable circumstances for people using the road such as taking care of trees and the areas overlooking the road .

2.7 Maintenance Importance and its Economic Advantages:

First: Reducing road deterioration.

Figure (2.2), which is a typical performance curve, clarifies the relation between the road condition and its age. When the road gets older due to its use by vehicles, gradual deterioration appears. Generally, the deterioration average depends on many factors including weather condition, paving quality, underlying soil strength, traffic volume, and axial loads for the trade vehicles allowed to be used on the road.

Neglecting routine maintenance, such as neglecting the small cracks on the paving surface, will cause gaps through which water penetrates into base layer or the layer below the base one. This matter leads to sink in certain parts of the roads. Subsequently, repeated periodic maintenance is required whose cost are (10) times more than the routine maintenance cost .

From what has been mentioned above, it can be recognized the economic advantage in implementing maintenance works properly. It can be concluded, also, that the roads after a certain period of time require maintenance works. Improvements and rehabilitation of roads are important to keep them as a longer period as possible in service as shown in Figure (2.3), which indicates a new circle of the road's life.

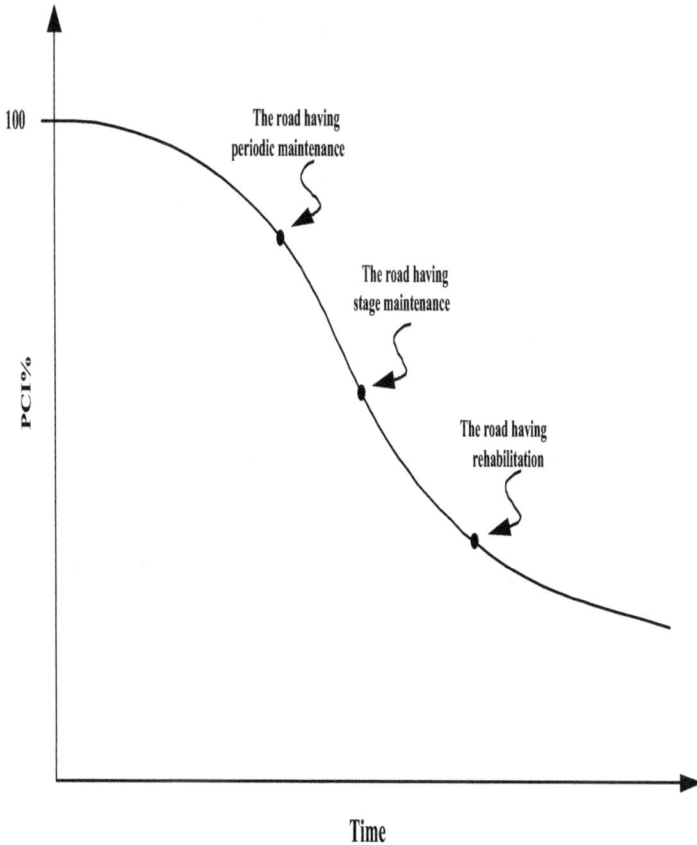

Figure (2.2) Performance Curve Typical .

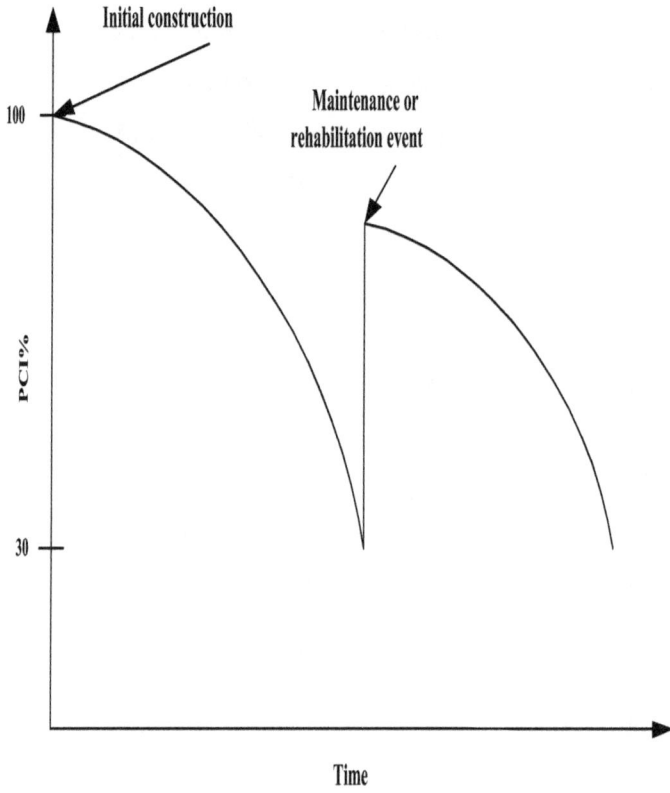

Figure (2.3) Concept of projecting pavement performance using PCI.

Second: Lowering Vehicle Operating Cost.

Driving vehicles in damaged and narrow roads will increase the operating cost, because of more fuel consumption. Statistical studies, in this respect, were published lately in the United States of America indicate that the waste of fuel is about (16.4) billion gallons per year because of using damaged roads.

In addition to that, damaged roads will increase the vehicles wheels consumption, as well as, the suspension system is consumed in a shorter time.

Subsequently, everyone in the United State of America has to pay a hidden tax which is about (136) dollars every year.

Thus, keeping roads new by maintenance will provide a comfortable and easy driving and reduce the cost borne by road users. Subsequently, a great benefit for the national economy will be achieved by paying attention to maintenance and improving the road to lower the vehicles operating cost .

Third: Keeping the Road Open.

The third justification for implementing maintenance works is to keep the road open to ensure providing continuous service for residential area, industrial and agricultural facilities surrounding the road. Blocking the road either because of the sand hills or floods in rain seasons will also create series of social and economic consequences. In one of the African countries, 40% of vehicles are unable reach their destinations due to the bad condition of the roads in rain seasons. The matter, which led to close a number of factories and plants for several months, as well as, delivering their production to distribution centers. Additionally, the agricultural areas at that were affected for not being able to get the fertilizers or delivering their products to the markets.

3 HIGHWAY DISTRESSES

3.1 Introduction

The deterioration of a pavement can be distinguished by various external markers or indicators, and often these indicators reveal the probable causes of the failure. The discussions of problems regarded to pavement distress generally depend on whether the pavement has a flexible or rigid surface type. Generally, the various types of distresses fall into one of the following categories:

(A) Rutting:

Description: A rut is a surface depression in the wheel paths. Pavement uplift many occur along the sides of the rut, but, in many instances ruts are noticeable only after a rainfall when the paths are filled with water. Rutting stems from a permanent deformation in any of the pavement layers or subgrades, usually, caused by consolidated or lateral movement of the materials due to traffic load. Significant rutting may lead to major structural failure of the pavement.

(B) Depressions:

Description: Depressions are localized pavement surface areas with elevations slightly lower than those of the surrounding pavement. In many instances, light depressions are not noticeable until after a rain, when ponding water creates a "bird bath" area; on dry pavement. Depressions can be spotted by looking for strains caused by ponding water. Depressions are created by settlement of the foundation soil or they are the results of improper construction. Depressions cause some roughness, and when deep enough or filled with water, may cause hydroplaning. Sags, unlike depressions, are abrupt drops in elevation.

(C) Corrugation:

Description: Corrugation (also known as wash boarding) is series of closely spaces ridges and valleys (ripples) that occurs at fairly regular intervals, usually less than 10 ft (3 m) along the pavement. The ridges are perpendicular to the traffic direction. This type of distress is usually caused by traffic action combined with an unstable pavement surface or base. If bumps occur in series of less than 10 ft due to any cause, the distress is considered corrugation.

(D) Potholes:

Description: Potholes are small-usually less than 3 ft (0.9 m) in diameter shaped depressions in the pavement surface. They, generally, have sharp edges and vertical sides near the top of the hole. Their growth is accelerated by free moisture collection inside the hole. Potholes are produced when traffic abrades are small pieces of the pavement surface. The pavement, then, continues to disintegrate because of the poor surface mixtures, weak spots in the base or subgrade, or because it has reached a condition of high-severity alligator cracking. Potholes most often are structurally related distresses and should not be confused with raveling and weathering. When holes are created by high-severity alligator cracking, they should be identified as potholes not as weathering.

(E) Shoving and Pushing:

Description: Shoving is a permanent, longitudinal displacement of a localized area of the pavement surface caused by traffic loading. When traffic pushes against the pavement, it produces a short and abrupt wave in the pavement surface. This distress normally occurs only in unstable liquid asphalt mix (cut back or emulsion) pavement.

(F) Weathering and Raveling:

Description: Weathering and raveling are the wearing away of the pavement surface due to a loss of asphalt or tar binder and dislodged aggregate particles. These distresses indicate that either the asphalt binder has hardened appreciably or a poor-quality mixture is present. In addition, raveling may be caused by certain types of traffic, e.g., tracked vehicles. Softening of the surface and dislodging of the aggregates due to oil spillage are also included under raveling.

(G) Polished Aggregate:

Description: This distress is caused by repeated traffic applications. When the aggregate in the surface becomes smooth to the touch, adhesion with vehicle tires is considerably reduced. When the portion of aggregate extending above the surface becomes small, the pavement texture does not significantly contribute to reducing vehicle speed. Polished aggregate should be counted when close examination reveals that the aggregate extending above the asphalt is negligible, and the surface aggregate is smooth to the touch. This type of distress is indicated when the number on a skid resistance test is low or has dropped significantly from a previous rating.

(H) Bleeding or Excess Asphalt:

Description: Bleeding is a film of bituminous material on the pavement surface that creates a shiny, glasslike, reflecting surface that usually becomes quite sticky. Bleeding is caused by excessive asphalt cement or tars in the mix, excess applications of bituminous sealant, and low air void content. It occurs when asphalt fills the voids of the mix during hot weather and then expands on to the pavement surface. Since the bleeding process is not reversible during cold weather, asphalt or tar will accumulate on the surface.

(I) Bumps and Sage:

Description: Bumps are small, localized, upward displacements of the pavement surface. They are different from shores in the sense that shores are caused by unstable pavements. Bumps, on the other hand, can be caused by several factors, including:

a) Buckling or bulging of underlying PCC slab in overlay over PCC pavements.
b) Frost heave (ice, lens growth)
c) Infiltration and buildup of material in a crack in combination with traffic loading "sometimes called a tenting".

Sage are small, abrupt, downward displacements of the pavement surface. Distribution and displacement that occur over large areas of the pavement surface, cause large and/or long displacements in the pavement are called a swelling.

(J) Swell:

Description: swell is characterized by an upward bulge in the pavement's surface-along, gradual wave more than loft (3m) long. Swelling can be accompanied by surface cracking and this distress is usually caused by frost action in the sub grade or by swelling soil.

(K) Cracks:

There are two main types of cracks as shown in Figure (2.1) they are transverse cracks and longitudinal cracks.

(L) Transverses cracks:

Transverse crack extends across the pavement at approximately right angles to the pavement centerline or direction of lay down. These types of cracks are not usually loads associated.

(M) Longitudinal Cracks:

It is parallel to the pavement's centerline or lay down direction. They may be caused by:

a) A poorly constructed paving lane joint.

b) Shrinkage of the AC surface due to low temperatures or hardening of the asphalt and/or daily-temperature cycling.

c) Reflective cracks caused by cracking beneath the surface cause, including cracks in PCC slabs (but not PCC joints).

There are several types of longitudinal cracks.

(M-1) *Edge joint cracks:*

Description: edge cracks are parallel to and usually within (2ft or 0.6m) of the outer edge of the pavement. This distress is accelerated by traffic loading and can be caused by frost-weakened base or sub grade the area between the cracks and pavement edge, which is classified as raveled if it breaks up (sometimes to the extent that pieces are removed).

(M-2) *Lane joint cracks:*

These cracks are between road lanes and caused by the pound between the distribution's lanes is weakness or the traffic loading. Traffic loading on the mentioned lane is different from the other lane.

(M-3) *Alligator cracks:*

Alligator or fatigue cracking is a series interconnecting cracks caused by fatigue failure of the asphalt concrete surface under repeated traffic loading. Cracking begins at the bottom of the asphalt surface (or stabilized base) where tensile stress and strain are the highest under a wheel load. These cracks resemble an alligator's skin. Alligator cracking is considered as a major structural distress and is usually accompanied by rutting.

(M-4) *Shrinkage cracks:*

Block cracks are interconnected cracks that divide the pavement into approximately rectangular pieces. The blocks may range in size from approximately (1 by 1 ft) (0.3 by 0.3m) to (10 by 10 ft) (3 by 3m). Block cracking is caused mainly by shrinkage of the asphalt concrete and daily temperature cycling (which results in daily stress/strain cycling), It is not load-associated. Block cracking usually indicates that the asphalt has hardened significantly. Block cracking normally occurs over a large portion of the pavement area, but sometimes will occur only in non-traffic area. This type of distress differs from alligator cracking, in that alligator cracks from similar, many-sided pieces with sharp angles. Also, unlike block, alligator cracks are caused by repeated traffic loading, and are therefore found only in traffic area (i.e. wheel paths).

(M-5) *Slippage cracking:*

Slippage cracks are crescent a half-moon shaped cracks. They are produced when braking or turning wheels cause the pavement surface to slide or deform. This distress usually occurs when there is a low-strength surface mix or poor bond between the surface and the next layer of the pavement structure.

(M-6) *Reflection cracks:*

Cracks that appear on a surface or overlay are caused by joints or crack in underlining. They may also be described as longitudinal or transverse in relation to the road centerline.

4 MAINTENANCE OPERATIONS

4.1 Introduction

Maintenance, as defined earlier, involves keeping facilities as near the "constructed or reconstructed" condition as possible depending on circumstances. This kind of maintenance calls for periodic or almost immediate correction of unfavorable or unacceptable situations. For the brief discussions that follows, maintenance is categorized into surface, shoulder and approach, road side and drainage, snow and ice control, bridge and traffic service

4.2 Surface Maintenance:

Day to day assurance of roadway surface integrity is a primary responsibility of maintenance. Roadway surface includes the travel way and shoulders. Additional requirements for shoulder maintenance are presented in item (B) and about one-half of the state highway maintenance money return goes for routine care of the roadway surface.

Travel ways are structured of two parts, described as:

(1) Pavement structure is made up of layers described as sub-base, base coarse and surface coarse, and

(2) Subgrade, which is the graded portion of a highway (roadbed) prepared as a foundation for the travel way pavement structure and shoulders.

Surface courses may be untreated or treated with admixtures or may be paved. Pavement structure is usually described as either flexible or rigid for bituminous type and Portland cement concrete type respectively.

Maintenance requirements for roadway surface are intended to offset the effects of edge, wear, and various type of distress. The type of distress and the maintenance method employed are, largely, influenced by the type of pavement structure and materials used.

The range of variables that affect the life of a roadway surface can be extensive. Temperature change, moisture content, structural material, design and the type and amount of traffic are some of the factors that influence the length of time before a roadway will require resurfacing or other maintenance. The quality of maintenance is, of course, a considerable factor in preserving the lifetime of a pavement surface and its serviceability during that period. Quality of maintenance is based upon a combination of the right material and their proper and timely used.

Maintenance programs for pavement surfaces when asphaltic materials are used should be divided into two categories based on the technical requirements, involved in selecting the right materials and employing the best available methods. The first category includes operations such as patching, crack filling and use of surface treatments such as fog and sand seals. The second category includes resurfacing operations that involve precise determination of material characteristics and quantities to assure proper bonding.

Under the second category, the technology required to produce an effective surface by slurry seal or overlay should not be underestimated. Taking into consideration the time and materials that can be invested it is better to obtain good assurances that an effective product will result. Factors that affect resurfacing are, type of roadway, shape and condition, amount of grade and quality of aggregates ambient air and material temperatures; degree of surface moisture and humidity; condition and preparation of surfaces to be treated; traffic volume; equipment and method to be used; and grades of asphalt and aggregate and their relation to other materials, methods and carrying time. Determinations by materials laboratory can prevent losses when sizeable amounts of paving materials are used. Many surface failures are due to either improper construction or design. A knowledge of the cause of these failures is essential to proper maintenance and repair. Failures caused by design and those due to increasingly heavy traffic should refer to design engineers for correction construction projects. Surface failures that are attributed to poor construction usually require increasing amounts of corrective action by maintenance over the life span of the pavement.

4.3 Shoulders and Approaches Maintenance:

Maintenance requirements of shoulders and approaches are influenced by their design, usage, condition and structural materials. Shoulders are parallel and contiguous to the travel way to which they provide lateral support. They should be suitable to provide emergency parking or serve as deceleration lanes.

The proper shape of shoulders to prevent collection of water in pockets and to assure good drainage should be an important maintenance objective. Shoulder width types and cross slopes are determined by design standards on new highways. It should always maintain to proper grade.

Newly constructed shoulders should be checked frequently during the rainy season for damage, particularly for any slumping that will affect the shoulder's ability to provide lateral support to the travel way.

There are four types of shoulders:

(i) *Earth or sod:*

Earth or sodded shoulders should contain a sufficient amount of granular material for stability. Grass cover should be native and mostly volunteer, not requiring special irrigation. Earth or sod will tend to build up and should be bladed lightly to maintain proper cross slope drainage. Blading should be done mostly when the earth is moist, often enough to keep ruts filled and lightly enough to prevent damage to grass roots. Rats and soft spots appear after a rain or when ground frost has thawed and should get attention at these critical times. If shoulder material has been removed by erosion, it should be replaced.

In regions of heavy growth, use of a disc harrow or mechanical mixer in the early spring to break up roots and sprigs followed by blading is recommended. This will assist the normal growth of grass on the shoulders.

Sod shoulders should not be mowed shorter than $(3''-4'')$ or the effectiveness of the grass cover will be lost.

(ii) *Gravel or crushed stone:*

Gravel shoulders offer better support for traffic than sod and more desirable as an all-weather shoulder. When a smooth surface or proper slope is required, the work can best be accomplished by a motor grader. A good procedure is to blade the gravel to the edge of the pavement, and to move it back over the full width of the shoulder on the second pass. Care should be taken not to leave ridges or false ditches to trap water on the shoulder. Dirt or depress from the slope or ditch area should not be mixed with shoulder gravel.

Immediately following the blading of a gravel shoulder, it should be rolled to compact the material.

The application of dust palliatives on some shoulders may be advisable during the dry season particularly if there is a high traffic volume. Herbicides to prevent vegetation growth may be required. They should be applied prior to the growing season.

(iii) *Bituminous Paved:*

Paved shoulders offer better and safer facilities for acceleration and deceleration of vehicles. Paved shoulders should be palched and sealed as described in item (i) "roadway surfaces" sealing cracks and joints in shoulder pavement should be treated as prescribed for similar types of roadway surfaces. The weakest point of the paved shoulder and the place where most failures start in the joint between the travel way and the shoulder. Growth weeds and grass should be treated with herbicides prior to sealing cracks.

Shoulder gutters and curbs are constructed along the outside of the shoulder to collect and channel heavy runoff water into paved ditches or pipes at low points in the grades.

(iv) *Portland cement concrete:*

Portland cement concrete shoulders should receive the same maintenance as prescribed for rigid pavement.

4.4 Roadsides and Drainage Maintenance:

Roadsides are the areas between the outside edges of the shoulders and right-of-way boundaries. Unpaved median area between inside shoulders of divided highways and areas within interchanges are included roadsides should be maintained as nearly as practicable in a condition that supports management's intent or the configuration to which the roadside will subsequently mature and in a manner that contributed to safety, appearance, convenience and pleasure of the public and the preservation and protection of the roadway.

The largest budget allocations for roadside maintenance will be spent for vegetation control on the many acres of roadside right-of-way containing field grass, brushes, and native trees.

Highways are designed with drainage systems to keep water from damaging the structure by controlling or directing the free flow of water over, under or adjacent to the roadway. The drainage system of a modern highway is designed by an engineer who has a knowledge of hydraulics and highway drainage practices and who takes into account a considerable number of factors including: the amount and rate of the rainfall; natural drainage features; water table; contour of the land; effects of hydraulic pressure; etc..

Maintenance functions include:

a)   Keeping watercourses free from accumulations of dirt, debris, vegetation and other obstructions.

b)   Correcting malfunctioning parts of system, e.g. erosion and breaks or shifts in structures, waterways, conduits and pavings.

c)   Anticipating problems and making limited changes or modifications.

4.5 Bridge Maintenance:

Bridge maintenance is affected by bridge design, location, age, and use. Bridges are critical to highway operation and expensive (they represent and estimate 30% of highway inventory costs).

Most bridge maintenance is of a specialized nature. On structures having exposed steelwork, cleaning by sandblasting, flame, or other means followed by repainting usually represent the biggest maintenance item.

Deck joints may extrude or become filled with dirt so that cleaning and resealing is necessary. On occasion, vehicles out of control strike handrails or other appurtenances and these must be repaired. If bridge decks become rough or slick, resurfacing is in order. In this instance surfacing weight may be a primary control on what is used. Remedial measures are sometimes required to correct serious scour around and under piers and abutments. Because bridge maintenance requires special skills, it is common practice among highway agencies to have travelling crews exclusively for bridgework. Painting and other specialty work is often carried out under contract.

Deterioration of concrete bridge decks under the combination of freezing and thawing and deicing salts has become a major maintenance item. Commonly, the problem begins when the salt penetrates to and corrodes the reinforcing, resulting in spalling of the overlay concrete. Correction may require removed of the concrete, cleaning the steel, and applying new material, possibly polymer concrete. Sometimes, sealants or overlays of asphaltic material are employed to provide further protection, such major repairs do not cost much but may cause serious delays and inconvenience to materials.

4.6 Traffic Surface Maintenance:

Traffic surface includes such continuing functions as stripping, sign repair and maintenance of street and highway lights and signals. Generally, these are performed by special cross of the highway agency although streetlight and even traffic signals maintenance may be turned over to the local utility company.

At times, signal systems malfunction because of power or other failures. And as traffic-control devices and systems become more sophisticated and complex so the probability of breakdown increase. At the same time, the consequences of break down can be very severe. To keep such systems operating and to correct break down quickly calls for highly skilled personnel. Who has experience with such systems accumulates, they will become more reliable. Traffic service, also, include coping with emergencies. For example, during heavy storms maintenance personnel will be on patrol to try to keep the roads open, sign or barricade washes out and rescue stranded materials.

4.7 Snow and Ice Control:

Snow removal is the major winter maintenance problem in affected areas. To solve this properly requires careful organization and advanced training removal operations should start soon after snow begins to fall.

REFERENCES

Acra, I., Fil, E., Fim, U.N.E., Fih, E., and Mrtp, I., **"Maintenance standard for roads in rural country."**, The Highway Engineer, The Journal of the Institution of Highway Engineers, Vol. 31, No. 3, March, 1974, p.p. 20-25.

America Association of State Highways and Transportation Officials, **"AASHTO maintenance manual."**, 1st Edition, 1976.

Faiq, M.S., **"The Use of Artificial Neural Network for Estimation Total Cost of Highway Construction Project"**, Ph.D. Thesis, Baghdad University, Iraq, 2009.

Haas, R. and Hudson, W.R., **"Pavement management system."**, McGraw-Hill Book Company, 1978.

Ismail, E.A., **"Highway intersections with alternative priority rules"**, Ph.D. Thesis, University of Bradford, England, 1989.

Lahey Computer Systems, **"Extended memory operating environment, user manual"**, Lahey Computer Systems, Inc., USA, 1988.

Oglesby, C.H. and Hicks, R,G., **"Highway engineering."**, 4th Edition, John Wiley & Sons. Inc., 1982.

Robinson, R., **"Road maintenance and management for developing countries."**, The Journal of the institution of highway and transportation, Vol. 33, No. 6, June, 1986, p.p. 8-12.

Shahine, M.Y. and Walther, J.A., **"Pavement maintenance for roads and streets using the PAVER system."**, U.S. Army Construction Engineering Research Laboratory (USACERL), Technical Report m-90/05, USA, July, 1990.

TRB, **"Traffic flow theory: A monograph"**, Transportation research board, Special Report 165, 1975.

U.N. Economic Commission for Africa, **"Road maintenance handbook."**, Vol. III, (Maintenance for paved roads), **Sited by (4)**.

ABOUT THE AUTHOR

Assist Prof. Dr. Faiq M. Sarhan Al-Zwainy

Place & date of Birth, Iraq – Baghdad , 1974

<u>**Qualifications:**</u>

- Ph.D. Civil engineering, Construction Management, Baghdad University–Iraq, 2009.

- M.Sc. Civil engineering, Construction Management, AL-Mustansryia University– Iraq, 2000.

- B.Sc. In Civil engineering, AL-Mustansryia University–Iraq, 1996.

<u>Bodies and Associations</u>

- Iraqi engineers union, Membership number is: 82026. in July / 1996

- The national society of engineering inspection and civil protection, Membership number is: 334 in September/2016

- Member of PMI(Project Management Institute), *https://my.pmi.org/*

<u>Academically experience:</u>

- Assist Prof. Dr.: Civil Eng. Dept. post graduated/ Al-Nahrain University, 2015- now.

- Assist Prof. Dr.: Civil Eng. Dept. under graduated/ Al-Nahrain University 2009 -now.

- A member of the committees discusses Master and doctoral dissertations in the specialty Project Management / University of Technology and the University of Baghdad since 2010 and now.

Google   **Scholar:**
https://scholar.google.com/citations?user=FeXQQ7AAAAAJ&hl=en
orcid.org/0000-0002-9948-6594
Scopus Author ID: 55347693000
ResearcherID: G-9229-2016
Personal Web-Site: http://www.drfaiq.blogspot.com
Communication: Mobile Phone. +964 7703925506, +964 7806567255
Mail:-faiq.al-zwainy@eng.nahrainuniv.edu.iq,
        faiq_faiqmohmed@yahoo.com

Dr. Ibraheem A. Aidan

Place & date of Birth, Baghdad, Iraq, 1962.
Qualifications:
1) Ph.D. Construction project Management, University of Technology - Iraq, 2001.
2) M.Sc. Construction project Management, University of Technology -Iraq, 1996
3) B.Sc. In Civil engineering, University of Technology -Iraq, 1984.

Bodies and associations
1) Member of Iraqi engineers Union, (1985-now).
2) Member of Arab engineers Union, (2014-now).
3) The national society of engineering inspection and civil protection, Membership number is: 334 in September/2016

Experience:
(2013-2014) Dr.: Civil Eng. Dept. under graduated ,Alfarabi University College, Iraq
1) (2015-Now) Advisor of M.Sc. Students.
2) (2010-Now) member of dissertation discussions of M.Sc. students at Al-Anbar University and University of Technology, (UOT).
3) (2010-Now) Scientific evaluator M.Sc. and paper researches.
4) (2003-Now) Consultant in Consultation services Bureau in Engineering Collage at Al-Nahrain University
5) (2005-Now) Manager of Consultant Engineering Group. (CEG).
6) (2002-2003) Assistant Dean of Engineering Collage, Al-Nahrain University
7) (2000-2003) Instructor Dr.: Civil Eng. Dept. under graduated at University of Technology, (UOT).
8) (2002-2003) Dr.: Civil Eng. Dept. under graduated at Al-Nahrain University, Iraq
9) (2002-Now) Professional trainer in the consultation bureau conducting engineering courses and programs pertaining project management.
10) (1999-2002) Head of civil engineering department in Engineering Affairs Office. Baghdad.

Google   **Scholar:**
https://scholar.google.com/citations?user=EYRIGDkAAAAJ&hl=en
Communication: Mob. 009647711313282;
Mail: ibraheemaudan@yahoo.com